ST. GEORGE SCHOOL
LINN, MO 65051

NO LONGER PROPERTY OF

Patty Cake

ELIZABETH MOODY

Photographs by photographers of *The New York Times:*

Neal Boenzi, Don Hogan Charles, Samuel Falk,

Edward Hausner, Jack Manning, Lee Romero, Barton

Silverman *and* Robert Beach *and* George B. Schaller

Designed by Joan Stoliar

QUADRANGLE **NYT** The New York Times Book Co.

THOMAS JEFFERSON [JS] LIBRARY SYSTEM

214 Adams Street
Jefferson City, Missouri 65101

Copyright © 1974 by
Elizabeth Moody. All rights
reserved, including the right
to reproduce this book or
portions thereof in any form.
For information, address:
Quadrangle/The New York
Times Book Co., 10 East
53 Street, New York, New
York 10022. Manufactured in
the United States of America.
Published simultaneously in
Canada by Fitzhenry &
Whiteside, Ltd., Toronto.

Library of Congress Catalog
Card Number: 73-90185

International Standard Book
Number: 0-8129-0433-8

PHOTOGRAPH CREDITS:

New York Times photographers

SAMUEL FALK, Endpaper
(front)
NEAL BOENZI, ii, 54, 58, 63
(bottom)
DON HOGAN CHARLES,
Frontispiece, 66, 68, 70,
72, 73, 74, 75 (right)
LEE ROMERO, 4, 6, 9, 11, 13, 18
(right), 19 (2 photos),
32, 36 (2 photos), 37, 39,
41, 42, 75 (left), 76, 77,
79 (2 photos), 80 (left),
80 (right, 2nd from top),
82 (top), 82 (right), 83
(top), Endpaper (back)
EDWARD HAUSNER, 17, 24, 26,
27, 28, 29 (2 photos), 30,
31, 33, 34, 43, 53, 59, 60,
62, 63 (top), 65
JACK MANNING, 20, 21
BARTON SILVERMAN, 35
(2 photos), 44, 47
(2 photos), 48, 49
(2 photos), 51, 56, 57
(3 photos), 83 (bottom)

ROBERT BEACH, Central Park
Zoo staff, 7, 12
(2 photos), 15, 18 (left),
22 (3 photos), 23 (2
photos), 80 (right top
and 2 bottom), 81, 82
(left, 2 photos)

GEORGE B. SCHALLER (B.
Coleman, Inc.), viii

ACKNOWLEDGMENTS

The information on gorillas in the wild is
obtained for the most part from the
fascinating scientific studies of George B.
Schaller, American zoologist and member
of the African Primate Expedition of
1959–1960. In his books, *The Mountain
Gorilla* (Univ. of Chicago Press, 1963) and
The Year of the Gorilla (Univ. of Chicago
Press, 1964; reprinted by Ballantine Books,
1964) he describes close-range observations
of the animals in the mountain forests of
East Central Africa—unique reports in that
he was able to almost live *among* the gorilla
tribes for many weeks as they roamed
their natural habitat.

My gratitude to Deirdre Carmody whose
excellent reporting on the gorilla family in
The New York Times inspired me to visit
them—and eventually to write this account.

I also wish to express thanks to the zoo
keepers who talked with me and patiently
answered questions, particularly Raul Ortiz
and Richard Regano.

"Devoted as I was from boyhood to the cause of the protection of animal life, it is a special joy to me that the universal ethic of *reverence for life* shows the sympathy with animals which is so often represented as sentimentality, to be a duty which no thinking man can escape."

—ALBERT SCHWEITZER
Out of My Life and Thought

229240

FOREWORD

As the ancient story goes, the Lord of
Creation made the animals so that man
would not be alone. He gave man dominion
over birds and beasts—"and brought them
unto Adam to see what he would call them."

So man was given the privilege of naming
the animals and of contemplating, along
with the Creator, the beauty of animal nature:
the varied forms, the grace and power,
the mystery of their being. With the dominion
and the privilege also came the responsibility,
of which mankind must ever be reminded.

Does Albert Schweitzer's universal ethic,
reverence for life, allow the caging of wild
animals in zoos? This is a continuing debate.
Schweitzer did not approve of menageries,
but he did of course believe in the
enlightenment of man, and worked all his
life toward that end. Would he have
approved of modern zoos?

Scientifically administered zoos *would* seem
to be permissible—they are not merely to
entertain and amuse but to increase our
understanding. Will observing the wild
animals help people to realize the necessity
for conserving the fast-dwindling wild
habitats of the world? Will zoo visitors see
more clearly the urgent need for protecting
the life still living free in these green
places? Such enlightenment of man—the
most dangerous animal—would ultimately
benefit all life on the endangered planet.

But zoos of the future (the near future)
must be designed to include *within* the
animal enclosures some living vegetation
and some earth underfoot as well as sky
above—and enough space to move in. These,
as well as the provision of good keepers and
proper diets, are rights owed the wild things.

E. M.

Gorillas in the Wild

GORILLA (gorilla-gorilla)

DESPITE THEIR FIERCE APPEARANCE GORILLAS
ARE ACTUALLY SHY AND PEACEFUL ANIMALS;
THEY BEAT THEIR CHESTS ONLY TO SCARE
AWAY SUCH ANIMALS AS LEOPARDS AND MAN.

—Sign on gorilla cage, Central Park Zoo

They are quiet, peaceful creatures living in groups—sometimes there are four or five, sometimes as many as thirty in a colony. They are nomads, moving at a leisurely pace through their home range, eating, resting, eating again and sleeping. Theirs is a society with definite and unchallenged ranks of dominance, with a powerful silver-backed male as leader. Age and experience as well as physical strength are factors determining which male becomes head of the troop.

The leader seems to ignore, or at least tolerate, the occasional amorous approaches the females initiate toward younger black-backed males in the group. This attitude, inherent in gorilla nature, contributes to the balance and repose of group living. There are signs of affectionate behavior toward the leader, and between young and old.

Silent for the most part, the big male can give a bloodcurdling roar when threatened. He has a ritual act of rising on his hind legs and beating a loud tattoo on his chest with both hands in a display of bravado to intimidate foes, and probably to relieve tensions of fear and anger when excited. These terrible roars, or screams, of high intensity are described by scientist George Schaller as "probably among the most explosive sounds in nature, consisting of but a single violent tone. . . ." When an enemy approaches—man being the prime threat—the leader may accompany his scream with a lunging, swiping charge. This is said to be mostly bluff, though few have stood their ground to prove it.

The two types of African gorillas—mountain and lowland—have only minor variations in appearance. There is evidence that in past ages the gorilla existed as one continuous population from West to Central Africa but during long periods of drought the forest receded, depleting the intervening population. Today more than 600 miles separate the two groups.

At about sunrise the gorilla tribe gets off to a slow start. Some awaken, yawn and sit up, only to lie back down again. When the leader rises and moves off in a certain direction, the others watch, then move after him. Soon they spread out to forage among the herbs and vines, sitting down to strip and munch the bamboo stalks and to select the wild celery and other plants in the abundant undergrowth.

After two or three hours of foraging, the leader stops to rest and the others follow suit. Most build crude nests on the ground by bending and breaking the vegetation around them into cuplike shapes. Some make platforms of bent branches in the low crotches of trees. Others merely lie down or sit and doze where they are. Infants sleep with their mothers, but all the rest have individual nests.

Several hours later they are again on the move, feeding with many pauses, and much sitting and looking about. The youngsters romp and play together, climbing, exploring, but never straying far. The group is mostly silent but when scattered in the thick undergrowth the apes grunt repeatedly, probably to keep track of each other. They spend most of their time on the ground but climb trees, cautiously, to sample leaves or bark and to look around.

At nightfall they build new nests (they never use the same ones twice) and languidly settle down for some thirteen hours of sleep. It is a lazy, unexciting existence for the most part, and all very satisfactory in civil gorilla society.

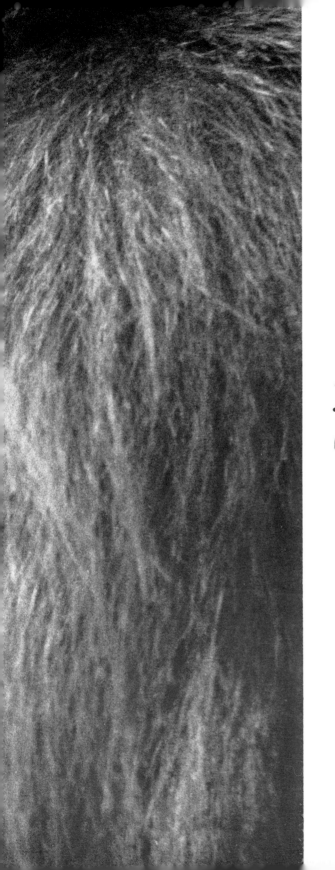

Patty Cake
On Scene

Ever since her birth at the
Central Park Zoo in New York,
Patty Cake, the baby gorilla
seemed destined for the limelight.

In the first place, she came as a surprise.
No one knew her mother Lulu was pregnant
—not even the devoted zoo attendants who
saw her daily.

It happened on Sunday, September 3, 1972,
when Lulu's favorite keeper, Richard
Regano, first got the news directly from
Lulu herself. Gently tugging his arm for
attention, she held up a furry black ball for
him to see. Then clutching her baby close,
Lulu retreated to a sunny corner and sat
down to await the reaction. It wasn't long
in coming. The story spread quickly and
other keepers and Sunday visitors rushed
to the cage; the Director of the Zoo arrived,
followed not long after by reporters and
photographers.

This was a first for Central Park.

Gorillas are an endangered species and few
are born in captivity. The great primates, in
many ways nearest to man among the
beasts, evoke a strange feeling of kinship
and wonder. That day Patty Cake made
news in papers across the land.

Lulu proved herself
an excellent mother.

She often showed affection by holding Patty Cake against her cheek, gently fondling and kissing her. At times she would roll over on her back, happily kicking her legs and holding the baby up to admire.

Lulu had naturally and without difficulty fed the baby at her breast. Patty Cake seemed relaxed and satisfied while feeding in the warm, protective closeness of her gentle mother. Gorillas are a gentle breed in spite of their awe-inspiring strength and appearance.

The experts were pleased. Gorilla females have no great reputation as mothers. Especially when in captivity they will often neglect or reject their infants, making it necessary for humans to take over their care. Patty Cake, it was hoped, would prove one of the few such animals raised successfully in captivity by her own parents.

12

Big Kongo, the two-hundred
and sixty pound father, was in
the same cage when Patty Cake
was born.

He had quietly watched the birth. But soon
after the infant was discovered, he was lured
into an adjoining cage, to prevent
accidental injury to his tiny daughter.

Kongo showed no surprise at the unusual
event. His lack of visible response is typical
of the species. Gorillas have an outwardly
placid nature and are not easily aroused.
Scientists have called them introverts who
are usually calm and "self dependent." They
don't show the excitability or avid curiosity
of monkeys or chimpanzees. And gorillas
are for the most part a silent breed.

Though Kongo sometimes rears up to his
full height and beats his chest with gusto,
no one has heard him demonstrate the
terrible gorilla roar—as yet.

When Patty Cake was a month old, Kongo was allowed to rejoin his family.

There was some anxiety about how he would treat the baby. Would he be jealous or hostile? He soon put such fears to rest.

At first Kongo stared intently at Patty Cake. Then he reached out and gently touched her with one finger. Whenever this happened, Lulu, always near, quickly carried the baby away, giving her mate glares or beseeching looks. She trusted neither man nor beast.

But Lulu's initial fear and possessiveness began to lessen. She did seem to enjoy the attention shown her infant by the visiting crowds. Although at first she kept Patty Cake well in hand toward the rear of the cage, she would sometimes come forward and appear to smile between the bars, always holding the baby close against her.

For a while Kongo was willing to indulge Lulu's possessive feelings toward the infant.

But later he asserted himself a little more and would occasionally grab the baby and gently fondle her in spite of Lulu's agonized reaction.

An infant of this species is completely dependent on the mother for the first few months, and her watchfulness and care at the early stage of the baby's life are necessities for survival when in the wild.

The rate of development in gorillas in the early months is approximately twice as fast as that of humans. There is a marked change in the gorilla youngster at about three months of age: he shows a growing awareness of his surroundings and of other animals. He has learned to focus his eyes on an object and reaches out for things, but his grasp is still weak. He can't as yet climb or hang onto his mother's back. But he begins to crawl and explore. He now has upper and lower teeth and can sample vegetables and fruit.

When their daughter was three months old and showing more independence, Lulu and Kongo could sometimes be seen sitting together watching her crawl about, seeming to enjoy a quiet family communication.

18

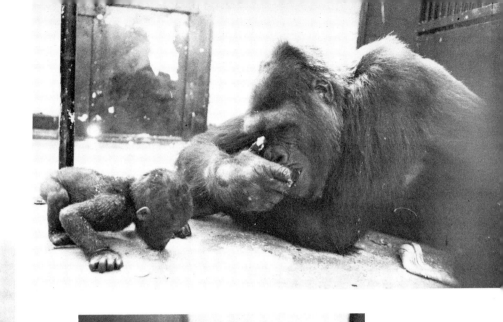

Kongo and Lulu had been raised together in the Central Park Zoo since they were one year old.

They are lowland gorillas and came from the Congo to Copenhagen to New York by way of animal collectors. The two were bought by the zoo in 1966 for $9,000 from Henry Trefflich, a well-known New York dealer, with the hope that they would eventually mate and reproduce the species. They had been good tempered, affectionate playmates from the beginning—too young when they arrived to pine for the green forest home of their birth.

On reaching maturity, Kongo and Lulu had naturally mated. It seemed a real love match. They were both seven years old when Patty Cake was born.

"Coming Out"

3 1331 00152 7060

It was not until she was six months old that Patty Cake had her real "coming out," as the Press called it.

It was set for the first Spring day in March when warmer weather began and the outdoor cages were opened. This meant the animals would enjoy more room and could go inside or outside as they pleased.

A sizeable crowd gathered to witness the debut. Lulu made it as dramatic as possible: she peeked out from under the drop gate of the inner cage, then retreated—several times. Finally, she cautiously emerged into the sunshine, holding her baby in one great hand.

For the first few minutes, Patty Cake hid shyly behind Lulu's big legs.

Onlookers were amused by Lulu's various protective holds—sometimes she would carry Patty Cake underarm like a football, sometimes against a shoulder with her huge leathery hand shielding the baby's head.

Patty Cake was precocious.

Her bright eyes showed interest in all around her. Though her outsized hands and feet still appeared too big for her to manage, she was becoming more agile every day. She crawled about and examined everything— including her mother's toes. Sometimes she could be heard making a soft panting, chuckling sound of contentment. Occasionally she whined a little. If she was startled by a sudden unfamiliar sound, she gave a high pitched screech.

Sometimes Patty Cake would grab her mother's leg and pull to a standing position. Then she would even totter a few steps on her hind legs.

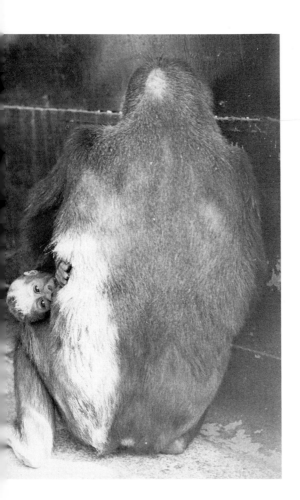

When Lulu turned her back on the world, Patty Cake often peered around her ample hip, fascinated by the passing parade.

She liked these gaping humans—such bright colors and funny sounds—and so friendly and admiring.

The small gorilla soon began to respond to her audience and show off with various antics. She could recognize her several keepers and would grab at their hands when they patted her on the head, and try to pat them back. She was especially drawn to the red hair of keeper Richard Berg.

No toys were given her while she was with her parents as the adult animals tore up such objects or threw them about and there was too much risk of an accident. Later on, however, she would be introduced to a little chimpanzee as an occasional playmate in a separate cage. They were then given balls to roll and ropes to swing on. Patty Cake liked to tinker with the big lock on the cage door, which to her was a shiny toy.

Of course the youngster took naturally to climbing. She was old enough now to grasp the bars firmly with both hands and feet and pull herself slowly skyward. When she neared the top, her mother would languidly reach up and pluck her down. She was not expert enough at this early age to make a safe descent—which Lulu seemed instinctively to understand.

She often crawled onto Lulu's rump and holding her tightly by the hair was jogged about, riding low-slung but never quite sliding off.

While her parents were munching fruit, Patty would sometimes pounce fiercely on discarded orange or banana peels, as if capturing some dangerous prey.

By now she had become one of
the most popular attractions
at the zoo.

A new star had been launched some two
miles north of Broadway and 42nd Street.

ST. GEORGE SCHOOL
LINN, MO 65051

Patty Cake's Ordeal

One Spring day as Keeper
Regano was making his
rounds he was startled by
a sudden scream.

He ran to investigate. It was Patty Cake!
She had suffered an accident.
The New York Times called it a family
misunderstanding. Kongo had reached
playfully for his daughter through the bars
in the adjoining cage, Lulu had grabbed
her away and accidently struck the baby's
arm against the bars, breaking the long bone
of the upper right arm.

Within a few minutes other keepers came
running to help. Lulu was tranquilized with
drugged darts and the baby taken from her.
Dr. Edward Garner, veterinarian from the
New York Medical College of Flower Fifth
Avenue Hospital hurried to the scene. He
ordered the little animal rushed to the
hospital for treatment. John Fitzgerald,
the Zoo Director, carefully carried her out
to the waiting car.

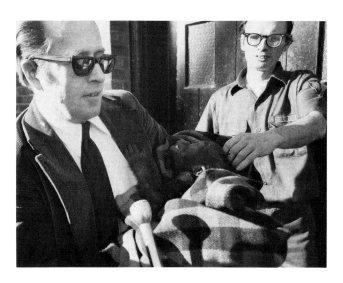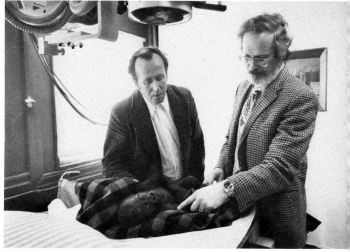

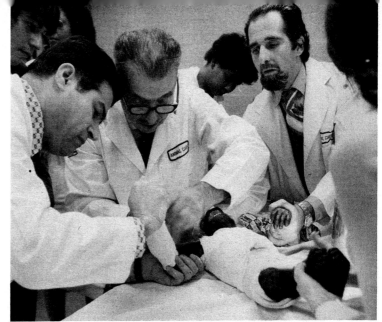

Dr. Montoya, a pediatrician, examined Patty Cake, who had been given a drug to ease her pain.

The operation performed by Dr. Michéle took an hour; afterwards she rested comfortably in an isolation room. She was soon alert again and the doctors said she was responding well to treatment although she was somewhat annoyed by the cast on her arm. John Fitzgerald was keeping an eye on the tiny patient. "Once in a while she'll come out with a little plaintive 'ooooo'," he said, "but she doesn't seem to be in pain and she'll soon find some way to maneuver."

Not only thousands of zoo visitors and others who read about her in the newspapers, but many scientists—zoologists, specialists in animal behavior and breeding—had by now become much interested in the gorilla family of Central Park. It would be a sad and disappointing loss if the baby was crippled or did not survive.

Dr. Garner, the veterinarian
who had been Patty Cake's doctor
since birth, sat up with her
all through the first night,
feeding her baby food of
chicken and pears.

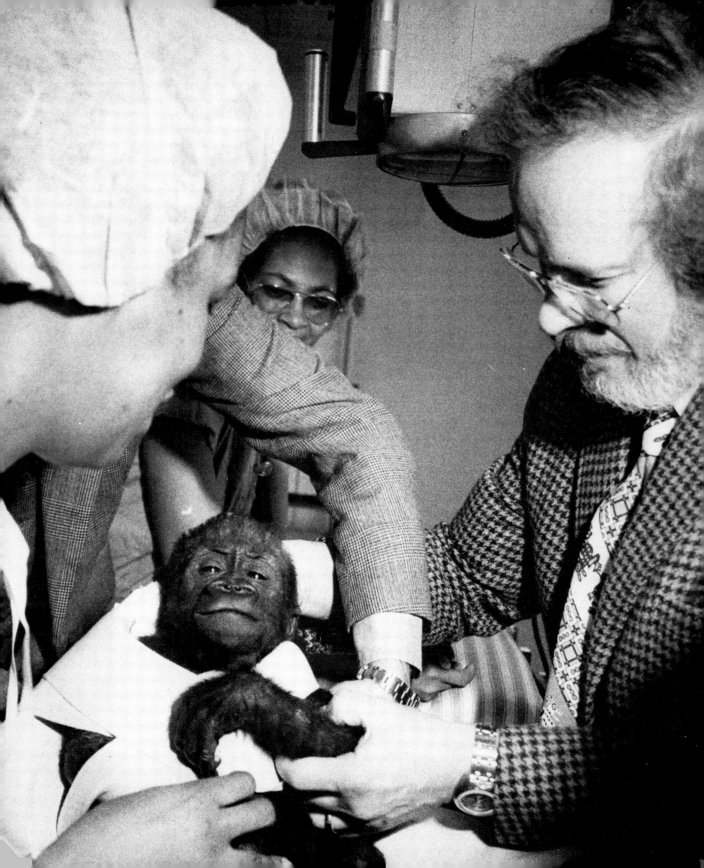

It was feared that if Patty Cake was returned to Lulu before her arm was entirely healed her mother might tear off the cast.

It would be months perhaps before the parent gorillas would see their offspring again. For some days Lulu seemed depressed and lethargic. It was clear that she was grieving.

Deirdre Carmody reported in the *Times*:
> . . . Yesterday, the parents seemed saddened by the loss of the lively little Patty Cake. They clung to each other, stroking each other's arms, hugging and running their hands across each other's face.
> Lulu had a small cut over one eye and Kongo kept peering at it, touching it with his finger and licking it solicitously. . . .

Many people worried about the gorilla family who had showed such obvious affection for each other.

It was decided that the baby would stay in the animal hospital at the Bronx Zoo during her convalescence as Central Park had no such facility.

For a a while Patty Cake slept in a plexiglass crib. The scientific staff was keeping close watch on her progress.

The Bronx Zoo had a nursery where she would have constant human supervision. Mrs. Sherry King took overall charge as she was experienced with anthropoid apes and was already raising two little gorillas, four and eight months old. She was bottle feeding them a milk formula like that given human infants, with a supplementary fruit and vegetable diet.

Patty Cake was assigned a special nurse, Mrs. Caroline Atkinson, and the two became very good friends. Patty Cake was gaining weight and becoming her lively self again. A few weeks after her arm sling was removed, and a month or so after the accident, Patty Cake was allowed outdoors to play. At first she seemed a little dubious but soon she began to enjoy herself.

Sometimes the three youngsters, Hodari, Patty Cake and Mopie romped together on the bars and swings. They wrestled and chased each other and appeared to be playing follow-the-leader. They were dressed in diapers as it made less clean-up problems for the handlers.

The little ape had entirely recovered.

Now almost nine months old, she weighed thirteen pounds, a gain of four and a half pounds since her accident. The Bronx Zoo had said she was undernourished on arrival, but babies that are breast-fed (whether human or animal) are usually a bit smaller than those who are bottle-fed. This of course does not mean that they are any less healthy. At the time of the accident, Lulu was still nursing Patty Cake, though she was also taking some solid foods.

The youngster was beginning to climb and explore. Her arms were becoming strong and supple. She could almost do a chin-up on the play bar, and liked to hang by one arm and then the other. How she did enjoy being swung back and forth by her nurse!

Patty Cake drew many fans to the Bronx Zoo where she was put on public view from time to time.

Patty Cake was thriving in the Bronx Zoo.

She was happy with her human caretakers and got along well with the other young apes in their frequent play periods. But did she remember—or unconsciously miss—her natural parents? Was she developing an "identity problem" being raised by people and dressed occasionally in clothes?

There is a mysterious wisdom which adult animals impart to their young in close and secret ways still unrevealed to mankind. Was Patty Cake being deprived of something vital to her unique gorilla nature? Experiments with chimpanzees a few years ago had shown that when the infants were separated from their natural mothers they seemed normal until adolescence. Then most of them showed strong neurotic traits and later proved unable to adjust well to other chimpanzees, or to mate easily and reproduce.

Some two months after the accident, Richard M. Clurman, Administrator of Parks, Recreation and Cultural Affairs (including the Central Park Zoo) made it known that he felt it was time for Patty Cake's return.

But William G. Conway, General Director of the New York Zoological Society (including administration of the Bronx Zoo) questioned whether she could be properly raised in the Central Park facilities. And of course no one knew how Lulu and Kongo would react to their daughter's return after this lapse of time. It might be too much of a shock for the little animal if she was rejected, or possibly even injured, and had to be moved again to other surroundings.

So the dispute over her custody became serious. It was finally agreed by officials on both sides to call in an outside mediator, an impartial expert whose decision would be binding.

Dr. Donald D. Nadler, a psychologist and development biologist at the Yerkes Regional Primate Center in Atlanta, was selected for this difficult job. The Yerkes Center, connected with Emory University, has sixteen gorillas, the largest collection in the world.

Dr. Nadler flew to New York and began a careful investigation.

The case was of scientific interest since there are still many questions on the raising of this rare species in captivity. Importation of gorillas to the United States has been drastically reduced since 1969 under the Endangered Species Act. As Dr. Conway put it: "The guiding principle is to do what is best for Patty Cake because of her potential as a future breeder."

Dr. Nadler observed Patty Cake in the Bronx Zoo nursery.

62

He looked at Lulu and Kongo in their big double-cage in Central Park. He studied the situation for days and talked with all most closely concerned at both zoos, concentrating especially on the keepers. Richard Regano gave his observations and opinions to the scientist while his friend Lulu peered through the bars as if listening.

Dr. Nadler inspected and photographed the indoor and outdoor facilities at both zoos. Administrators, aides, even devoted zoo visitors, were interviewed for his final report. After several days he returned briefly to Atlanta to consult a veterinarian at Yerkes, announcing that he was reserving decision for a week or two.

On June 5 a formal news conference was called at the Bronx Zoo.

The decision was announced: Patty Cake should be returned to her parents.

In his report Dr. Nadler said: "... I feel that returning Patty Cake to her parents in the Central Park Zoo would provide her with the best opportunity to develop the normal social repertoire of a gorilla and, thereby improve the chance that she participate in the perpetuation of her species." This is particularly important for a gorilla, the scientist emphasized, since, as a species, it is in danger of extinction in its natural habitat.

Dr. Nadler recommended four important measures.
1. The barred area between cages where the baby's arm had been broken should be covered or replaced.
2. The stress on Patty Cake should be lessened by having her Central Park keeper feed her for a few days in the familiar Bronx Zoo surroundings.
3. Now that Lulu was no longer lactating, the baby's diet should be supplemented and her regular bottle formula continued.
4. Periodic health examinations should be given the small gorilla to insure that she remain in her present good condition.

The Central Park Zoo moved at once on Dr. Nadler's recommendations with Bronx Zoo officials cooperating in every way. Patty Cake was put into temporary quarters at the Park while final preparations were made for her return to Lulu and Kongo.

Reunion

Veronica Nelson, a young Central Park keeper, had been chosen to reintroduce the baby to her former surroundings.

She had experience as a veterinary nurse, and was well known to the gorilla family. She seemed to have a special gift for inspiring trust in the animals she cared for. She had adopted a tiny capuchin monkey whose mother had rejected it and often made her rounds with the little creature draped contentedly across her shoulders.

It had been planned that the baby would enter the empty cage before the mother was let in. The building was closed to the public and only a few people were allowed to witness the reunion in order to protect the animals during these first tense moments. There was a good deal of apprehension about the outcome.

A few days later the youngster was carried by Veronica Nelson, from the temporary quarters to the cage where she would be returned to her parents.

Led by Director John Fitzgerald, Veronica and the entourage of keepers and zoo aides moved through the Park. Patty Cake, wrapped in a blue blanket, was wide-eyed and alert. She looked and looked this way and that to be sure nothing interesting escaped her view.

Photographers snapped a few pictures on the way as Patty Cake by now was a famous New Yorker, sometimes referred to in print as "the little superstar," or "the Shirley Temple of the animal world."

Veronica sat very quietly in the cage while Patty Cake crawled about examining the place.

She discovered the open mesh door and began to climb. After about fifteen minutes, Veronica left her alone in the big enclosure.

Now Lulu was admitted to the adjoining cage.

She immediately peered through the bars
to where Patty Cake was crawling around.
Then suddenly Lulu screamed—a terrible
piercing cry. She ran from one end of the
wall to the other, screaming and screaming.
Patty Cake huddled with fright and began
to screech in high-pitched tones. Then the
door between the cages was opened and
the mother loped in.

But suddenly she stopped: instead of
rushing forward to pick up her baby she
stretched out her arms toward the tiny
animal in a gesture of instinctive control.
It was as if she knew the infant must be
reassured before she was touched, and must
be protected from further shock. Lulu
moved away, then toward the crouching
baby, then away again. Finally she reached
out and touched Patty Cake and bent
down and looked into her eyes. Patty Cake
stopped crying as Lulu, retreating, then
advancing, touched her again.

Then the mother put an arm around the
youngster, and holding her close climbed
onto a raised platform. Patty Cake clung
to her now and looked into her face. She
seemed calm and content as Lulu kissed her.

Half an hour later big Kongo was admitted
to the cage next door. He sat very still,
intently watching Lulu and Patty Cake
for some minutes. When the door was
opened he went quickly toward them. At
first Lulu moved away at his approach.
But finally she sat down and allowed him
to touch Patty Cake and to nuzzle her face.
The gorilla family was reunited at last.

Summer Visit

It's a warm, sunny morning in New York.

The Central Park Zoo is filled with leisurely visitors. The usual crowd is gathered in front of the outdoor cage where the gorilla family lives.

Lulu is reclining on the narrow platform, utterly relaxed, with one arm thrown back over her head and one foot raised and clutching a front bar. Cuddled on her chest is the baby, Patty Cake, sound asleep and hardly visible against the black fur. Some of the children call out in coaxing tones: "Patty Cake! Patty Cake! *Please* wake up!"

Suddenly Kongo bounds in from the low back entrance. His appearance always brings a gasp. He is huge, jet-black, powerful—in his prime. His face is calm and his whole bearing dignified—almost majestic. He lopes quickly forward and passing Lulu gives her what seems to be a playful poke. She makes no sound but immediately leaves the ledge, hugging the baby against her, and sits down in a corner. Kongo plumps beside her and they watch as Patty begins to bumble around, holding an orange peel in her mouth.

Richard Regano comes along hosing out the row of cages. At sight of the water, Lulu instantly puts her child under one arm and disappears into the back-stage refuge. Kongo lunges toward Regano who plays the water in a high stream so that some spray falls on the animal's back. Then playtime starts in earnest.

Kongo rushes inside and is hosed smartly on the rump as he disappears. Soon a crafty face peeps cautiously from the entrance but dodges back as more water comes that way. Suddenly he lumbers out and across the cage, ducking through the low door into the adjoining enclosure. In and out he runs, each time getting drenched as he goes across. Kongo, the keeper and the crowd are enjoying it immensely.

With a bang the back-drop falls shut and the big ape is trapped outside. He realizes it at once and rears on his hind legs, facing the spray. Now he is obviously taking his shower bath and liking it. He beats himself under the arms just like a human, and raises his chin to get his neck washed. He turns around to have his back showered, and turns again so that his stomach is tickled with the spray. As Regano moves away, the inner door is raised and Kongo makes a dripping exit.

Around noon keepers Luis Cerna and Raul Ortiz come with bottles for the whole family: milk for Patty Cake and grape juice for Lulu and Kongo. (The parents have their own bottles as they made such a fuss when only Patty Cake was provided one the first time.) The bottles are held between the bars while the animals drink; often Kongo just opens his big mouth and the juice is squirted in.

Later on, while the father lies in lazy siesta, the mother plays zestfully with Patty Cake. With one hand, Lulu rolls the baby over, pretending to bite her. The youngster staunchly defends herself, kicking, swatting, showing her teeth. She turns a few somersaults and makes fierce lunges at her mother. Then Lulu grabs and hugs her close as they smile together.

80